Editor April McCroskie
Language Consultant Prue Goodwin

Dr Gerald Legg holds a doctorate in zoology
from Manchester University. His current position is
biologist at the Booth Museum of Natural History
in Brighton.

Carolyn Scrace is a graduate of Brighton
College of Art, specialising in design and illustration.
She has worked in animation, advertising and
children's fiction. She is a major contributor to the
popular *Worldwise* series.

Prue Goodwin is a lecturer in Language
in Education, and director of INSET at the
Reading and Language Information Centre
at the University of Reading.

David Salariya was born in Dundee, Scotland,
where he studied illustration and printmaking,
concentrating on book design in his post-graduate
year. He has designed and created many new series
of children's books for publishers in the U.K.
and overseas.

Printed in Belgium

An SBC Book conceived, edited and designed by
The Salariya Book Company
25 Marlborough Place Brighton BN1 1UB

A CIP catalogue record for this book is available from
the British Library

ISBN 0 7496 2656 9
Dewey Classification 598.6

First published in Great Britain in 1997 by
Franklin Watts
96 Leonard Street
London
EC2A 4RH

Franklin Watts Australia
14 Mars Road
Lane Cove
NSW 2066

lifecycles

From Egg to Chicken

Written by Dr Gerald Legg
Illustrated by Carolyn Scrace

Created & Designed by David Salariya

W
FRANKLIN WATTS
NEW YORK • LONDON • SYDNEY

A chicken is a bird.
All birds have feathers and
two wings.
Chickens start life
inside an egg.
A chick hatches
from the egg
and grows into a chicken.
In this book you can
see this amazing
life cycle unfold.

The hen collects straw
to make a nest.
Then she will line the nest
with feathers to make
a soft bed.

The nest is
made of straw.

A hen is a
female chicken.

9

The hen lays her eggs in the nest.
She lays several eggs at a time.
Each egg holds and
protects a baby bird.
The egg also contains yolk,
which provides food
for the young chick.

A hen is a
female chicken.

The hen lays her eggs in the nest.
She lays several eggs at a time.
Each egg holds and
protects a baby bird.
The egg also contains yolk,
which provides food
for the young chick.

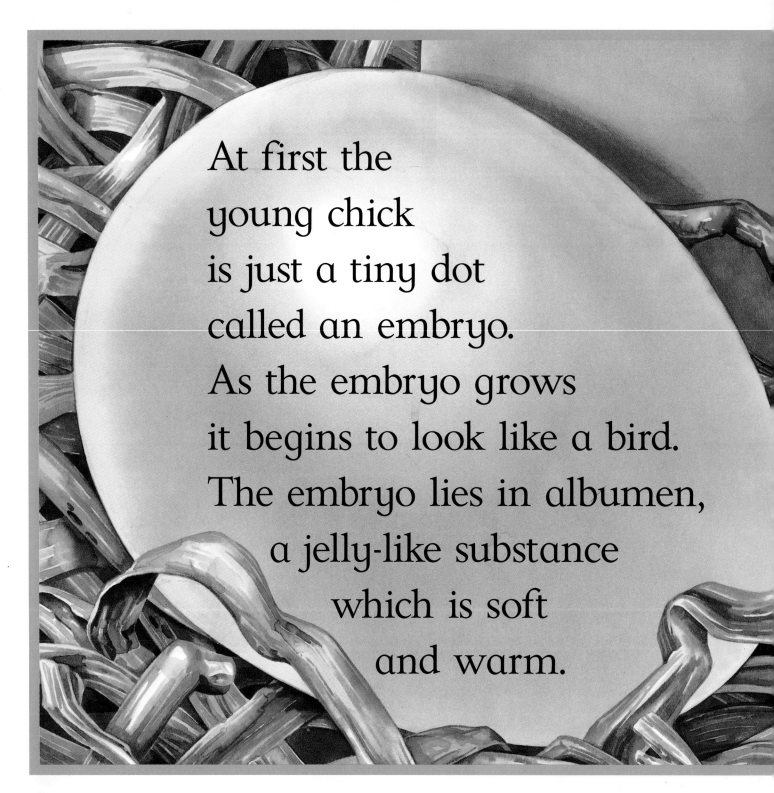

At first the
young chick
is just a tiny dot
called an embryo.
As the embryo grows
it begins to look like a bird.
The embryo lies in albumen,
a jelly-like substance
which is soft
and warm.

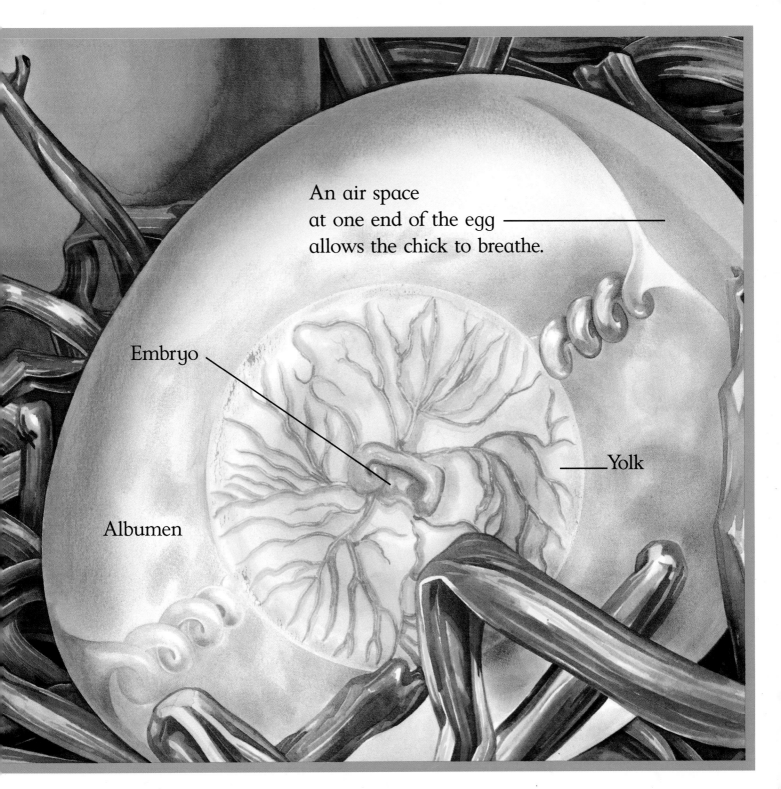

An air space
at one end of the egg
allows the chick to breathe.

Embryo

Yolk

Albumen

The hen is
a good mother.
She keeps her eggs
safe and warm
by sitting on them.
She surrounds them
with her soft feathers.
This is called brooding.

Every now and then,
the mother hen
turns the eggs
to keep them
warm all over.

Hens need to protect their eggs.
Birds, foxes, lizards, snakes
and humans all like to eat eggs.
To get at the food inside,
some animals bite
and peck at the eggshell.
Others smash it with stones.

Fox

After 21 days
the baby chick inside the egg
starts to make a cheeping noise.
This lets the mother hen
know that her chick
will hatch soon.
Then the eggshell
begins to crack.

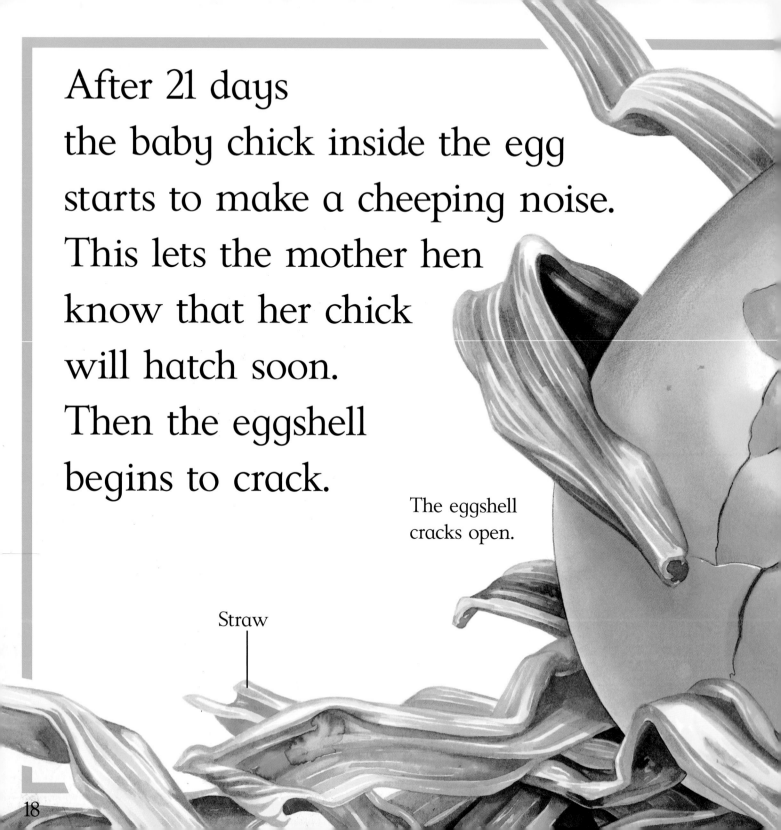

The eggshell
cracks open.

Straw

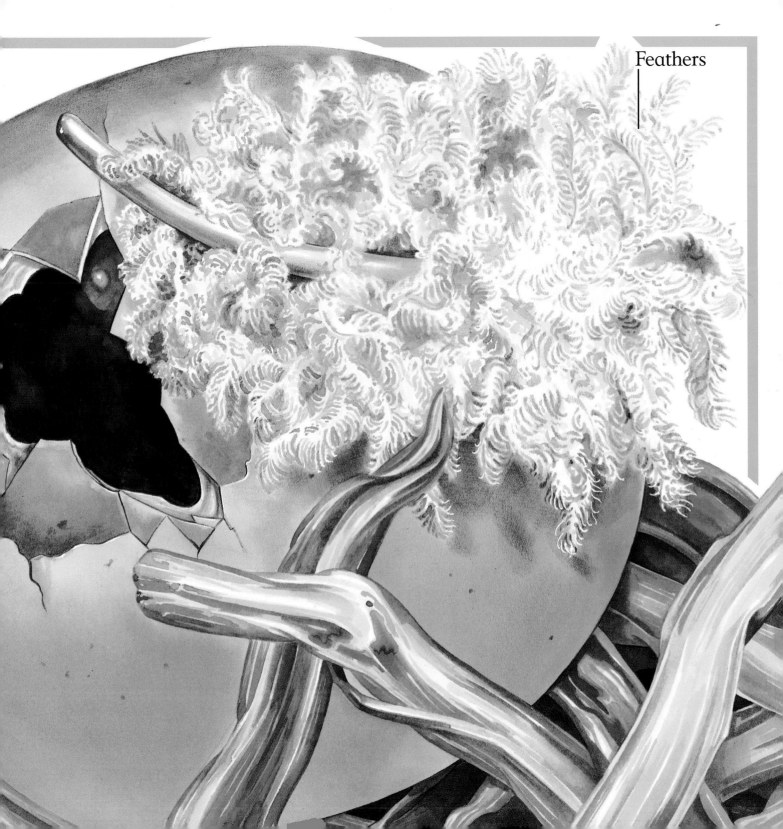

Feathers

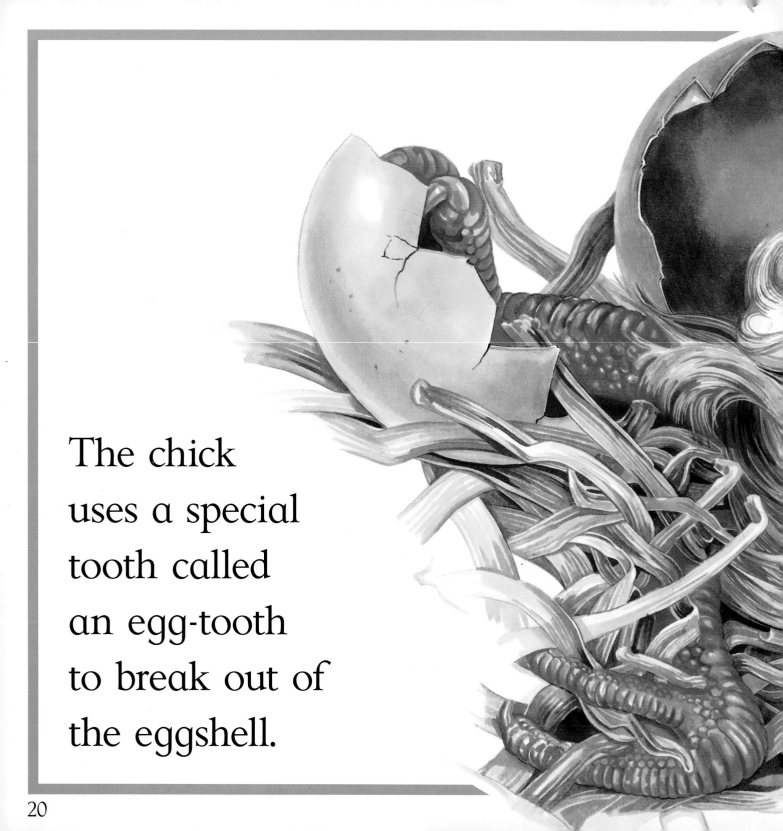

The chick
uses a special
tooth called
an egg-tooth
to break out of
the eggshell.

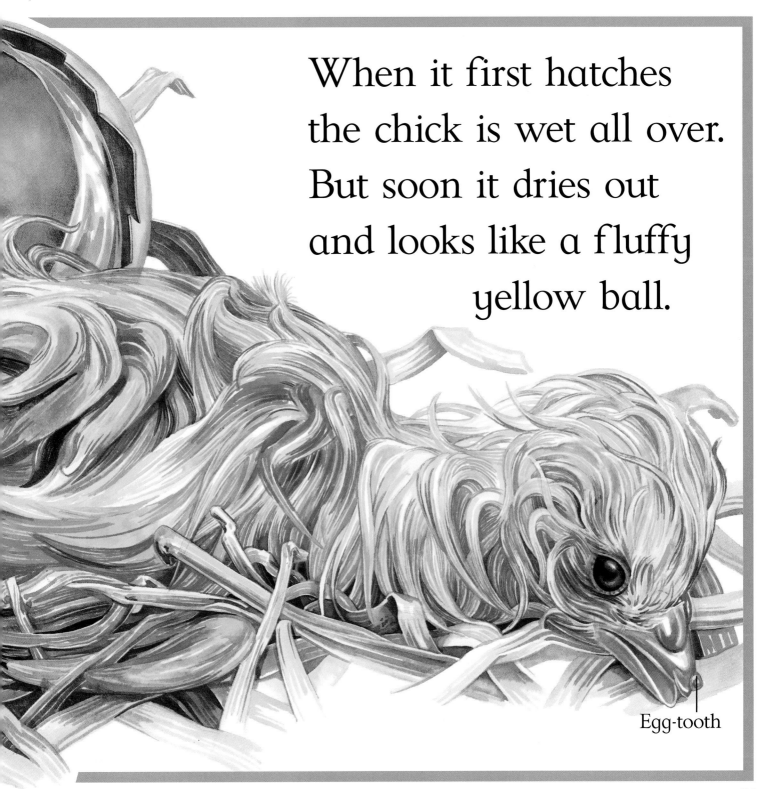

When it first hatches
the chick is wet all over.
But soon it dries out
and looks like a fluffy
yellow ball.

Egg-tooth

Chicks make
a cheeping noise
so that their mother
always knows
where they are.

The hen scratches
around for food.
The chicks follow her.
Older chicks are more brave
but they still stay near their mother.

As the chick gets older,
new feathers grow.

Comb

In a few months the chick will
look like its mother or father.
A female chicken is ready
to lay her first eggs when
she is about 25 weeks old.

This type of chicken is called a bantam. It has a bright red comb on its head.

Chicken facts

The American Poultry Association recognises 113 different breeds of chicken.

The domestic chicken has ancestors that live in the tropical forests of Asia, Africa and South America. These are called jungle fowl.

A new breed of chicken called a Frizzle has curly feathers.

Silkies are a breed of chicken that have hair-like feathers.

The largest-ever chicken weighed 10 kilograms and was of the White Sully breed.

An average chicken weighing 2.5 kilograms has a wingspan of around 86 centimetres.

The wandering albatross of the Antarctic has the biggest wingspan of all birds. Its wings stretch for 3.5 metres.

The growth of a Chick

In the pictures below you can see the way an egg grows into a chick. These pictures are the same size as a real egg and a real chick.

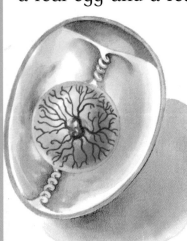
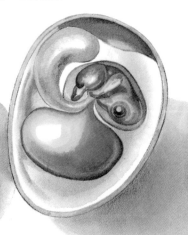
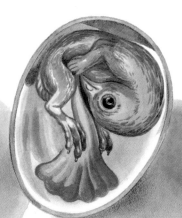

| 2 days | 7 days | 13 days | 16 days |

Eggs are strong. A typical chicken's egg will support a weight of 4 kilograms before breaking.

A good laying hen will produce 250-300 eggs a year.

Most eggs are oval, but owls lay eggs that are round.

The most eggs ever laid by a single chicken in a year was 371.

The largest-ever chicken egg was 31 centimetres long and had 5 yolks.

The heaviest-ever chicken egg weighed around 454 grams.

Ostriches from Africa lay the largest eggs. They weigh up to 3.9 kilograms and take 40 minutes to cook.

The smallest eggs are laid by hummingbirds. They weigh around 0.35 grams.

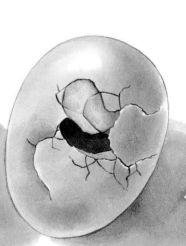

20 days

Newly hatched

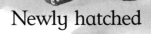

One-day old

Chicken words

Albumen
The white part of the egg which surrounds the growing chick.

Brooding
When a hen looks after her eggs by keeping them warm.

Cheeping
The noise that chicks make so that their mother knows where they are.

Comb
The bright red crest on top of a chicken's head.

Egg
Contains the baby bird.

Egg-tooth
A tiny tooth-like point on the tip of the beak of a new-born chick. The chick uses the egg-tooth to break through the eggshell.

Embryo
The early stage of a young animal before it can move and before it looks anything like its parents.

Feathers
The soft, light and often colourful covering of birds.

Index

Nest
A hollow place built or used by a bird as a home to rear its young.

Shell
The hard covering of an egg which protects the growing chick.

Yolk
The yellow part of an egg which is used as food by the growing chick.